BRONTOSAURUS IS BACK!

Stomp!

You're about to meet one of my extinct relatives: the awesome Apatosaurus!
Stomp!
Stomp!
Stomp!

BRONTOSAURUS IS BACK!

A Dinosaur's Name and Return to Fame

Sarah Albee

Illustrated by Chris Hsu

Charlesbridge

Behold! It's . . .

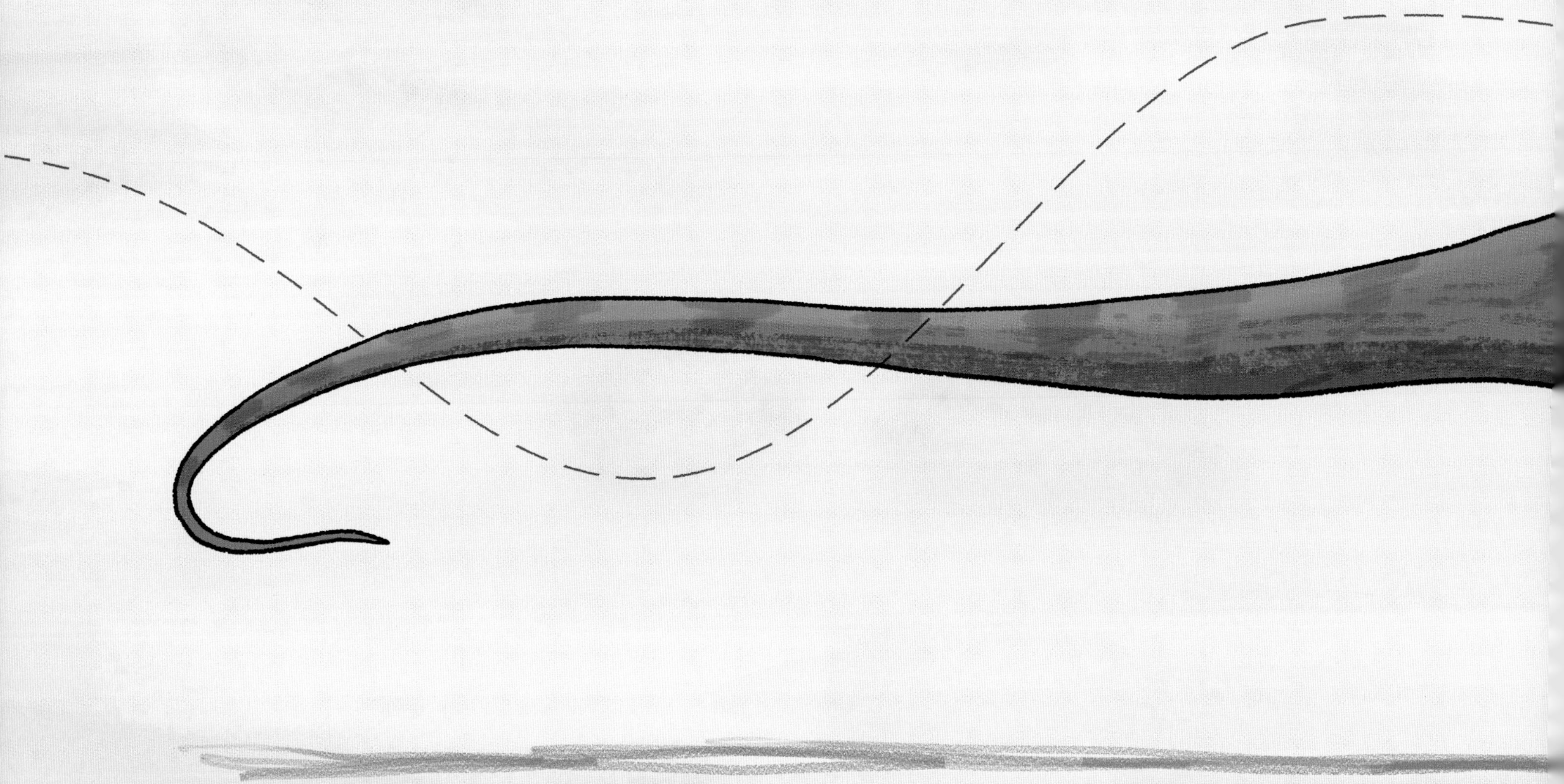

Brontosaurus!
Wait! Mistake! This is *Apatosaurus*. We're not supposed to call it *Brontosaurus*.

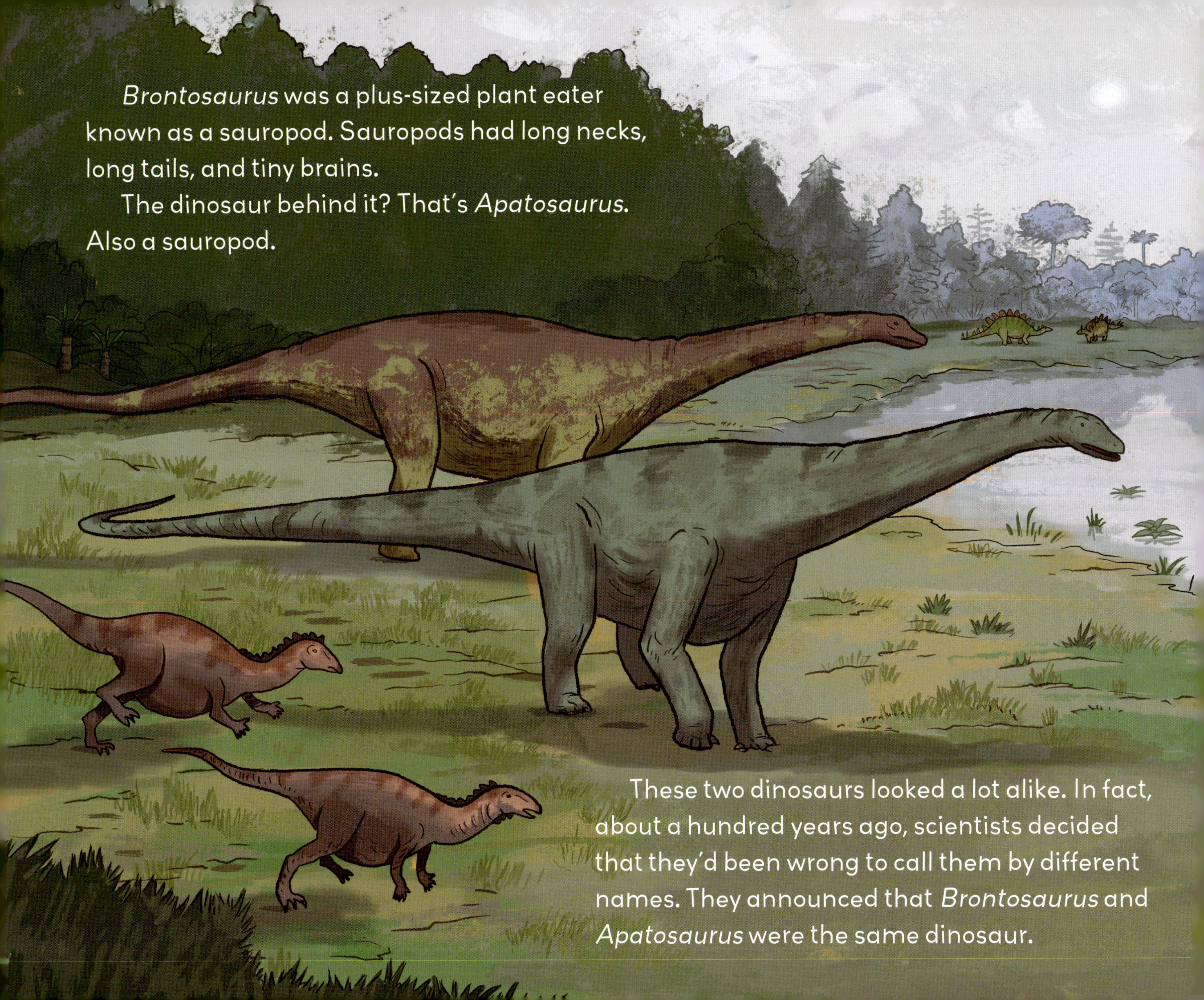

Brontosaurus was a plus-sized plant eater known as a sauropod. Sauropods had long necks, long tails, and tiny brains.

The dinosaur behind it? That's *Apatosaurus*. Also a sauropod.

These two dinosaurs looked a lot alike. In fact, about a hundred years ago, scientists decided that they'd been wrong to call them by different names. They announced that *Brontosaurus* and *Apatosaurus* were the same dinosaur.

Suddenly the name *Brontosaurus* became invalid. Defunct. Wrong.

But times have changed! Let's see how the confusion started.

Dino Times

Sauropods lived during what scientists call the Mesozoic Era (the Age of Dinosaurs). Scientists divide that era into three parts: Triassic, Jurassic, and Cretaceous periods. *Brontosaurus* and *Apatosaurus* lived in the Late Jurassic period, which ended about 145 million years ago.

In places such as ancient China, ancient Greece, and early North America, people had been stumbling over huge bones for thousands of years. Many cultures created myths about dragons and giants from the past.

In what eventually became the western United States, the Sioux people told stories about "thunder beasts" that lived long ago.

The Earliest Words

Scientists first used the word *dinosaur* in 1842. It means "fearfully great lizard" in Greek.

A paleontologist named Othniel C. Marsh learned about some large, fossilized bones from a Sioux chief named Red Cloud.

Marsh wanted to find those fossils.

Talk Like a Scientist

Fossils are the evidence of life that lived long ago. A paleontologist is a scientist who studies the fossils of ancient living things.

Marsh had a rival. His name was Edward D. Cope. Cope was also a paleontologist. Marsh and Cope became bitter enemies. They constantly competed with each other.

Both wanted to discover and name the most dinosaurs and other extinct animals. Were they good fossil finders? Yes. Were they sometimes slapdash? Also yes.

One day while digging out West, members of Marsh's team uncovered some fossilized bones. Even though some were in bits and pieces, it was clear they had belonged to something enormous. **Massive. Gigundo.** Could these be the remains of one of those "thunder beasts"? Perhaps a new species, or even a new genus?

By that time—the 1870s—the railroads had been built, so the men loaded the bones onto a train and sent them to Marsh, who was back East in Connecticut.

Bone by Bone

A genus is a group of closely related living things. *Brontosaurus* is a genus. A genus may include several species. A species is a group of even more closely related living things. *Brontosaurus excelsus* is a species.

Sure enough, Marsh concluded that the bones belonged to an unknown, super-sized dinosaur. He rolled up his sleeves and got to work examining them.

All Marsh had to work with were a few loose vertebrae and a fragment of a pelvis. But did that stop him from claiming that he had discovered a new genus? Of course not!

He named the new dinosaur *Apatosaurus*. That's Greek for "deceptive lizard."

Piece by Piece

Paleontologists rarely find complete skeletons. Often they have only fragments of bones to work with. So they reconstruct missing pieces based on what they know about closely related species.

Not long after that, Marsh received the fossils of a second specimen. After a rapid examination, he announced that he'd discovered another new dinosaur. He named it *Brontosaurus*, which means "thunder lizard." Way cooler name, right?

A few years later, Marsh's crew uncovered more *Brontosaurus* bones. But they didn't find a head.

No head? No problem! Marsh produced a picture of a *Brontosaurus* skeleton and guessed at the shape of its head based on a different, equally enormous sauropod. As usual he raced to publish his results.

Head Case

Dinosaur skulls are relatively fragile and often don't survive the fossilization process. When skulls *are* found, they're often detached from the skeleton, making it hard to know whose head is whose.

Meanwhile, Marsh's competitor Cope was also busy digging up dinosaur bones. In just over two years, Marsh and Cope found and named a *lot* of sauropods.

ATLANTOSAURUS!

APATOSAURUS!

CAMARASAURUS!

AMPHICOELIAS!

MORROSAURUS!

DIPLODOCUS!

The upside of this frenzy of fossil discoveries? The public became dino-dazzled. Museums everywhere wanted fossils to display. And sauropods were wildly popular.

The downside? Because Marsh and Cope were in such a hurry to outdo each other, they rushed to claim discoveries based on scant evidence and got a lot wrong. Those blunders perplexed many later generations of paleontologists.

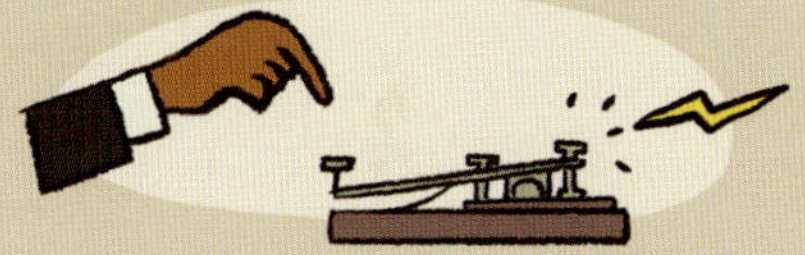

The Name Game

To claim a new discovery back then, all Marsh or Cope had to do was send a telegram to a scientific journal to announce that they had discovered something. According to the rules of scientific naming, the person who discovers a new species gets to name it. Historically the name was in Latin or Greek, although today dinosaur names come from many other languages.

And people kept digging up more and more sauropod fossils.

A few years after Marsh and Cope died, a scientist named Elmer S. Riggs made a careful comparison of some *Apatosaurus* and *Brontosaurus* fossils. His conclusion? Marsh was wrong. *Apatosaurus* and *Brontosaurus* were *the same genus.*

Marsh had named *Apatosaurus* first. So just like that, the name *Brontosaurus* ceased to exist. It was changed to *Apatosaurus*.

A Dinosaur Disappears

The rules of scientific naming specify that the first name assigned to a living thing in a science-based publication is the valid one.

But some museums ignored the updated science. They mounted what should then have been called an *Apatosaurus* skeleton and labeled it *Brontosaurus*. (People really liked that name.)

For the next hundred or so years, the name *Brontosaurus* stayed stuck in people's minds. Scientifically speaking, it was incorrect, but only paleontologists and dino-obsessed kids seemed to care.

Most museum visitors knew nothing about the name confusion. They just loved to gape at those enormous creatures from the past.

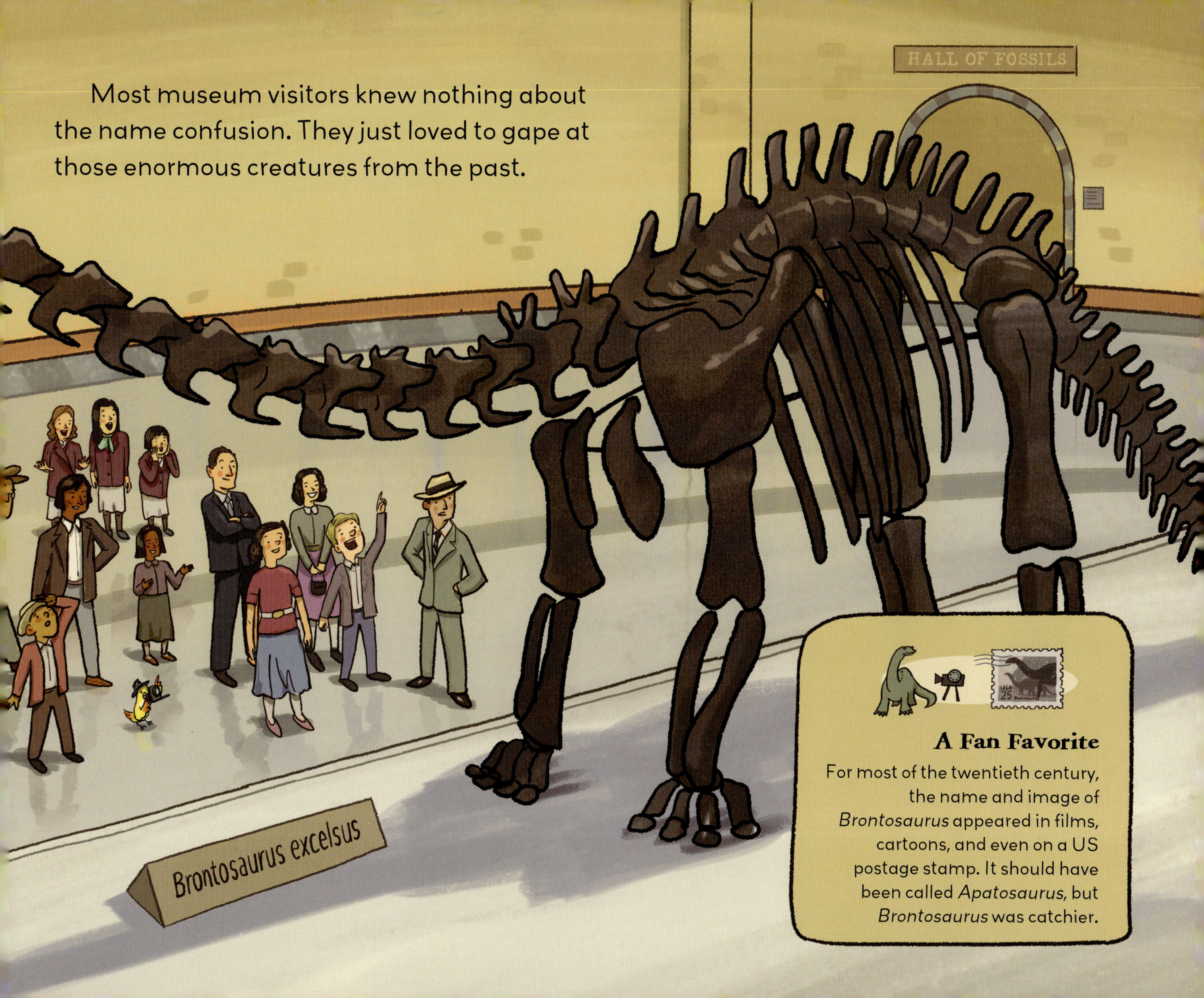

A Fan Favorite

For most of the twentieth century, the name and image of *Brontosaurus* appeared in films, cartoons, and even on a US postage stamp. It should have been called *Apatosaurus*, but *Brontosaurus* was catchier.

Apatosaurus? *Brontosaurus*? Whatever this sauropod's name was supposed to be, scientists still had no idea what its head looked like.

Was it snub-nosed like *Camarasaurus*?

Or long and horsey like *Diplodocus*?

But finally, in 1978, two scientists made a discovery. They located a skull that had been stored in a museum basement for nearly seventy years. After a careful examination, the two scientists determined that the skull belonged to *Apatosaurus*.

The head made headlines. Museums around the world removed the incorrect heads on their *Apatosaurus* skeletons and replaced them with more accurate versions.

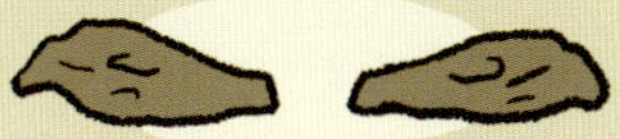

A Head Transplant

The skull the two scientists identified belonged to the species *Apatosaurus louisae*. Today we know that although *Apatosaurus* and *Brontosaurus* were different in many ways, their heads probably looked pretty similar.

Over the decades more and more—and *more*—sauropod fossils were discovered.

And then came an even bigger Bronto bombshell!

A group of scientists published a large sauropod study in 2015. Their conclusion? *Brontosaurus* and *Apatosaurus are* different dinosaurs!

Dig This!

The results in this 2015 study reversed what scientists believed to be true for more than one hundred years.

Today scientists still have lots of questions about *Brontosaurus* and *Apatosaurus*. They aren't certain what the two dinosaurs looked like when they were alive, because fossils can't tell a whole story. For starters, we don't know what color patterns their skin may have had. But studying skeletons *does* help scientists understand how the two dinosaurs differed.

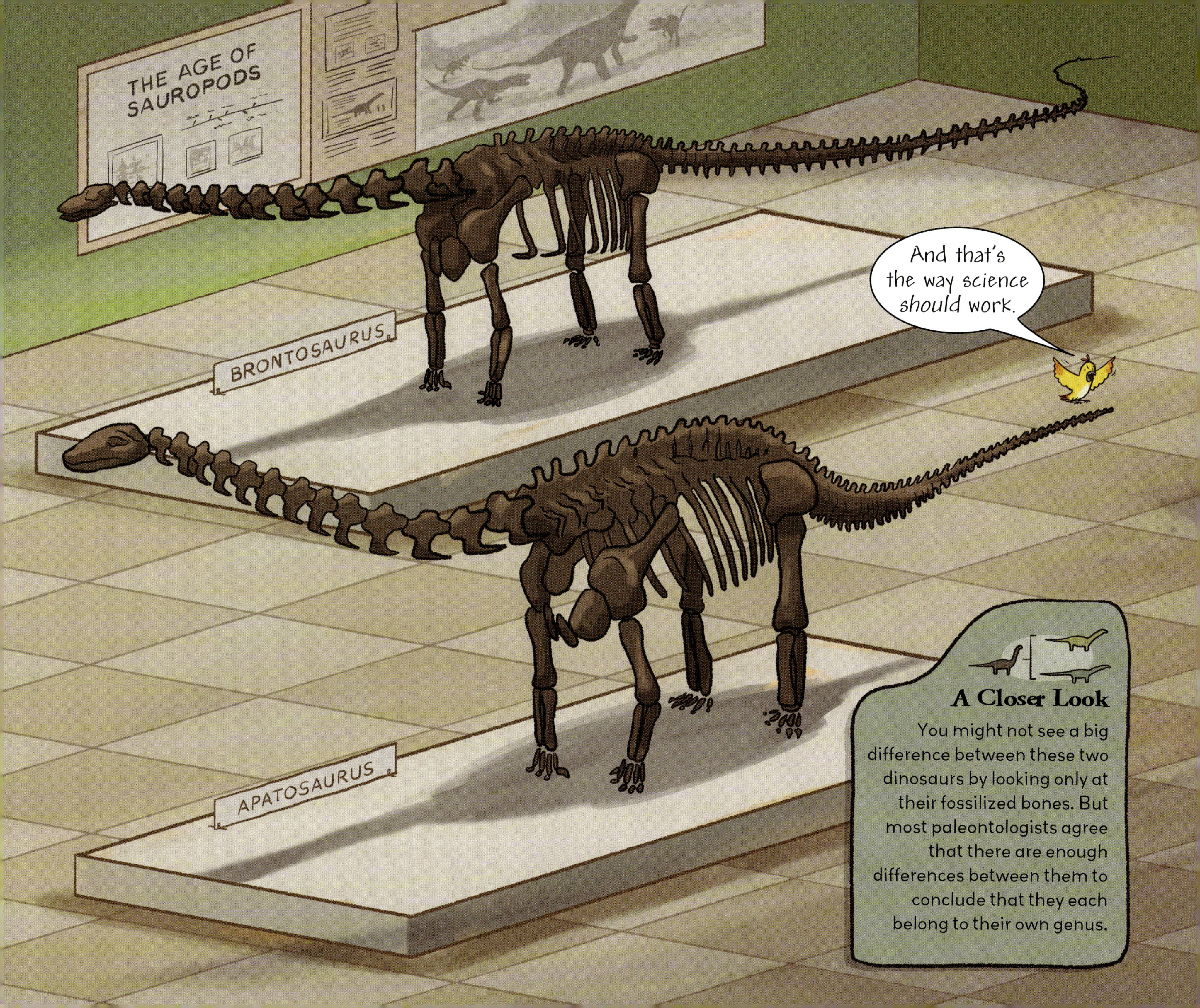

A Closer Look

You might not see a big difference between these two dinosaurs by looking only at their fossilized bones. But most paleontologists agree that there are enough differences between them to conclude that they each belong to their own genus.

And scientists have solved many mysteries about how these sauropods managed to keep their enormous bodies functioning.

Air sacs helped the animal breathe extra oxygen and may have kept its body from overheating.

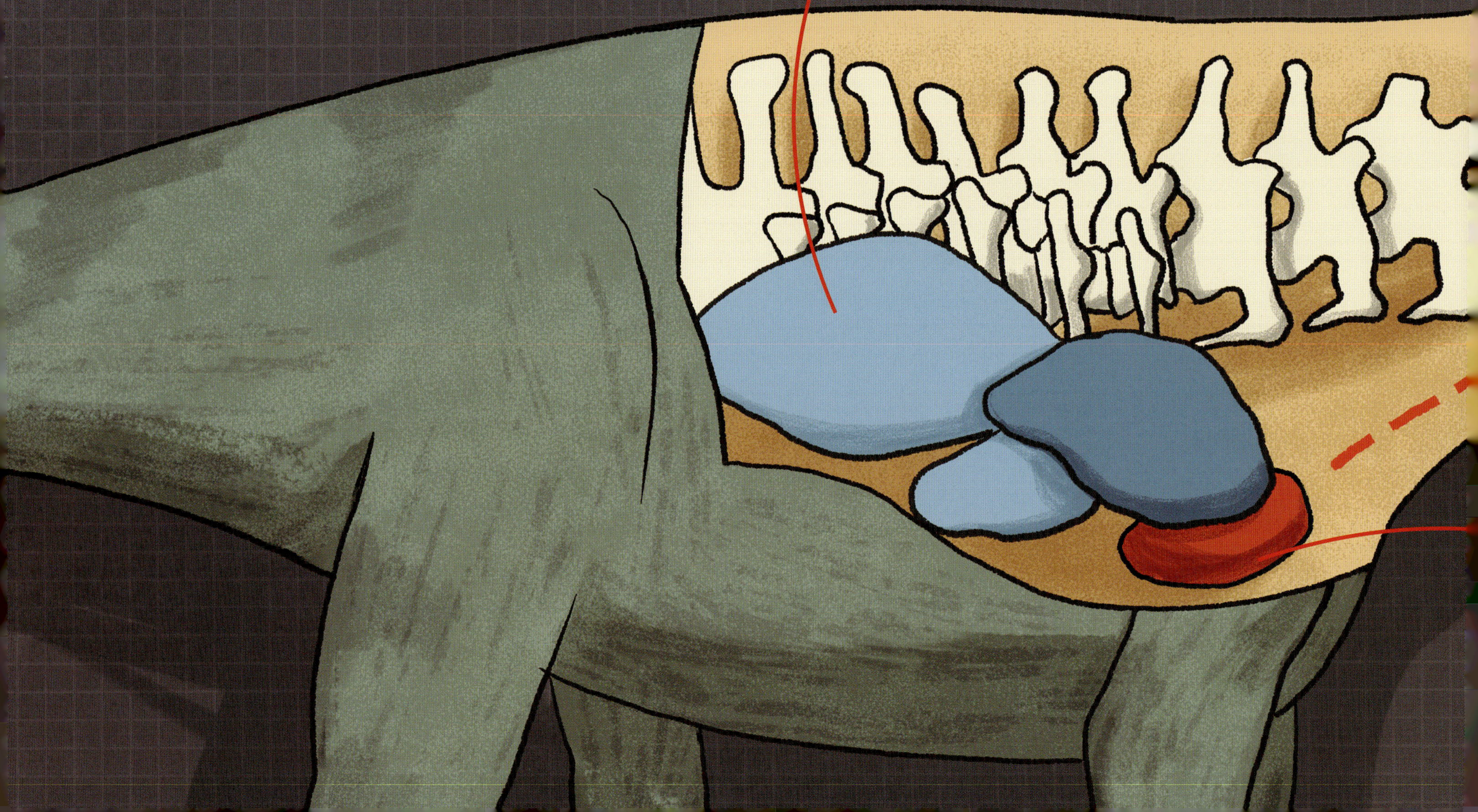

Spaces between bones helped lighten the animal's mass, and extra neck bones made the neck super flexible.

A huge heart pumped blood all the way up the neck to reach the brain, although scientists are still trying to figure out how that worked.

Baby Brontos

Because they were born small—about the size of a goose—life for these baby sauropods could be dangerous. They ran quickly for cover as soon as they hatched. If they avoided being eaten by predators, they grew big quickly. Scientists think they reached the size of a cow in just a few years. They reached full size in about twenty years.

To fuel their big bodies, *Brontosaurus* and *Apatosaurus* had to eat constantly. An adult tore off and swallowed about half a ton of plants each day.

Sauropods' humongous bodies scared off most predators. But if a meat eater did try to pick a fight? These sauropods knew how to defend themselves!

Pounding Plants

During the Jurassic Period there were no flowering plants. *Brontosaurus* and *Apatosaurus* ate vast amounts of ferns, conifers, horsetail plants, ginkgo trees, and cycads (sort of like modern sago palms). They had no chewing teeth, so they swallowed plants whole.

There's still so much to discover. Every time someone digs up a new sauropod bone, we learn a little more about these amazing animals.

But today we can say . . .

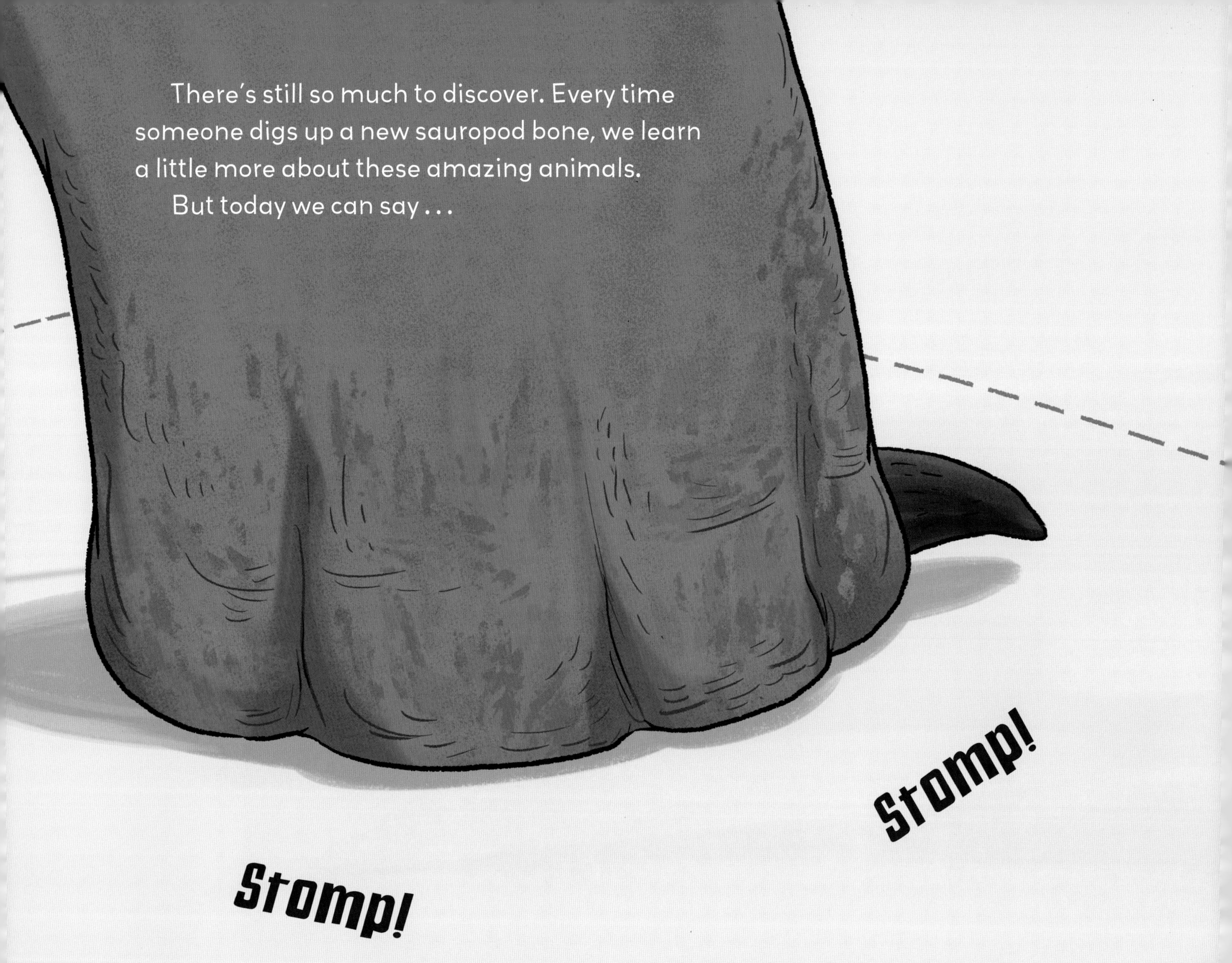

Brontosaurus is back!
And better than ever!
Stomp!

Timeline

154–145 million years ago: *Brontosaurus*, *Apatosaurus*, and other sauropods live in the late Jurassic Period. Many of these huge sauropods live in what is now the western United States.

1830s: Although huge bones were found in many places by many people for thousands of years, the 1830s mark the first scientifically recorded finding of a sauropod. Scientists call it *Cetiosaurus* ("whale lizard") because they believe these huge creatures lived in the water. (They didn't.)

1842: A scientist named Sir Richard Owen invents the word dinosaur. But he doesn't think *Cetiosaurus* was a dinosaur. (It was.)

1848: Most scientists conclude that *Cetiosaurus* was a type of dinosaur.

1877–79: Marsh and Cope discover many different species of sauropods. Marsh names *Apatosaurus ajax* based on just a few bones.

1878: Because most scientists now accept that these huge animals lived on land and not in the water, Marsh changes the broader name from cetiosaur ("whale lizard") to sauropod ("lizard foot").

1879: Marsh names a new species: *Brontosaurus excelsus*.

1883: Marsh's team finds another *Brontosaurus* skeleton, but because no skull is found, Marsh concocts what he thinks the head might have looked like based on a few fragments from a different site. (The head in Marsh's drawing resembles that of *Camarasaurus*, another sauropod dinosaur.)

1897: Cope dies.

1899: Marsh dies.

Early 1900s: More and more sauropod bones are found, and museums mount skeletons for display.

1903: A paleontologist at the Field Museum of Chicago named Elmer S. Riggs publishes an article that says *Apatosaurus* and *Brontosaurus* are the same animal. Because *Apatosaurus* was named first, the name *Brontosaurus* is no longer considered scientifically correct. Riggs's article doesn't get a lot of publicity.

1905: The American Museum of Natural History in New York City exhibits a skeleton based on bones collected in different parts of the American West and labels it *Brontosaurus*, ignoring Riggs's study. The public is fascinated by this enormous creature, and the name *Brontosaurus* sticks in people's minds even though it's not scientifically correct at the time.

Late-1960s/early-1970s: Although many paleontologists still think sauropods were too heavy to have lived on land, new evidence shows these giants walked upright. The absence of a tail drag (a mark along the ground made by a dragging tail) near fossilized footprints leads scientists to conclude that sauropods held their tails off the ground.

1978: Scientists John McIntosh and David Berman re-examine a head that had been found in the early 1900s and stored in the basement of the Carnegie Museum of Natural History in Pittsburgh. They publish an article that describes the head of *Apatosaurus* as more slender and horse-like than the snub-nosed, *Camarasaurus*-like skull many museums use for exhibits. A popular (but incorrect) myth springs up and persists for many decades that *Brontosaurus* is simply *Apatosaurus* with the wrong head.

Camarasaurus

Diplodocus

Apatosaurus

1979: The Carnegie Museum of Natural History replaces the *Camarasaurus*-like skull on its *Apatosaurus* skeleton with the proper one. A few years later, the Peabody Museum at Yale, the Field Museum in Chicago, and the American Museum of Natural History in New York do the same.

1995: The American Museum of Natural History changes the tail on its mounted skeleton. The tail is now held up in the air rather than dragging on the ground.

2011: An *Apatosaurus* skeleton with an intact head is finally found, in Utah. The finding confirms that the recent skull constructions are more accurate.

2015: A major study published by scientists Emanuel Tschopp, Octávio Mateus, and Roger B. J. Benson determines that *Apatosaurus* and *Brontosaurus* were different enough from each other to be classified as two different genera. After a careful comparison, the scientists find that although the living animals may have looked similar, there are significant differences in their necks, backs, shoulders, and foot bones. (Not all paleontologists agree, however, and the discussion is still ongoing.)

Today: Based on current evidence, scientists have named three species of *Brontosaurus* and two species of *Apatosaurus*.

New Evidence Leads to New Ways of Thinking

Science is a process. As new evidence is discovered, scientists can get a clearer picture of *Brontosaurus* and other sauropods.

What Scientists Used to Think	How Scientists Shifted Their Thinking Thanks to More Data
Sauropods crawled like lizards. →	Sauropods walked more like modern elephants.
They dragged their tails along the ground. →	They held their tails up.
To support their huge body, they had to live in water. →	They lived on land.
They were tubby and slow. →	They could move nimbly and even stand on their hind legs to reach higher into trees.
Their brains were tiny so they must have been dimwitted. →	Their brains were just the size they needed to be.

How Do We Know?

Paleontologists can tell a lot about a dinosaur from studying fossilized hard parts, such as bones and teeth, as well as fossilized soft stuff, such as skin, muscles, footprints, and even fossilized poop!

Fossilized teeth and poop can reveal the kind of food a dinosaur might have eaten. Both *Brontosaurus* and *Apatosaurus* had small, pencil-shaped teeth in the front of their mouth that were good for tearing plants. The absence of flat chewing teeth suggests these dinosaurs swallowed plants whole.

Leg bones can tell us how heavy a dinosaur might have been. Footprints can tell us how fast or slowly an animal moved. The absence of a tail drag tells us the dinosaur may have used its tail for balance. (Both *Brontosaurus* and *Apatosaurus* probably held their tails parallel to the ground.) Several side-by-side trackways can tell us the animals moved in herds. (*Brontosaurus* and *Apatosaurus* probably did.)

Scientists have even found fossilized eggs. Egg fossils can tell us about nesting behavior, the number of eggs a female dinosaur laid, and the size of dinosaur babies. The eggs of large sauropods were slightly smaller than a soccer ball.

What more do we want to find out?

Scientists today continue to learn about *Brontosaurus* and its close relatives. But there's still much more to discover.

What sorts of sounds did it make? What color was it? How did its heart pump blood all the way up its neck to the brain? How long did it live?

What else will we find out in the future about this amazing dinosaur? As more evidence is found, our knowledge of *Brontosaurus*, *Apatosaurus*, and other dinosaurs continues to improve.

A Note from the Illustrator

As an artist living in the twenty-first century, I did my best to illustrate what dinosaurs might have looked like based on the most current scientific theories. Because fossils don't reveal much about things like skin color and texture, a level of guesswork is always involved. One of the greatest mysteries of science will always be what dinosaurs actually looked like, and perhaps that mystery is one reason why humans will always continue to be fascinated with these prehistoric animals.

Select Bibliography (*denotes books for kids)

Berman, David S., and John S. McIntosh. "Skull and Relationship of the Upper Jurassic Sauropod *Apatosaurus*." *Bulletin of Carnegie Museum of Natural History* 8 (1978): 1–35.

Gould, Stephen Jay. *Bully for Brontosaurus: Reflections in Natural History*. New York: W. W. Norton, 1991.

Hallett, Mark, and Mathew J. Wedel. *The Sauropod Dinosaurs: Life in the Age of Giants*. Baltimore: Johns Hopkins University Press, 2016.

Jaffe, Mark. *The Gilded Dinosaur: The Fossil War Between E.D. Cope and O.C. Marsh and the Rise of American Science*. New York: Crown, 2000.

*Johnson, Rebecca L. *Battle of the Dinosaur Bones: Othniel Charles Marsh vs Edward Drinker Cope*. Minneapolis: Twenty-First Century Books, 2012.

Mayor, Adrienne. *Fossil Legends of the First Americans*. Princeton, NJ: Princeton University Press, 2005.

McIntosh, John S., and David S. Berman. "Description of the Palate and Lower Jaw of the Sauropod Dinosaur *Diplodocus* (Reptilia: Saurischia) with Remarks on the Nature of the Skull of *Apatosaurus*." *Journal of Paleontology* 49, no. 1 (January 1975): 187–199.

*Noyes, Deborah. *Tooth and Claw: The Dinosaur Wars*. New York: Viking, 2019.

Prothero, Donald R. "Is 'Brontosaurus' Back? Not so Fast!" Columbia University Press Blog, August 26, 2015. www.cupblog.org/2015/08/26/is-brontosaurus-back-not-so-fast-donald-prothero.

Riggs, Elmer S. "Structure and Relationships of Opisthocoelian Dinosaurs, Part I: Apatosaurus Marsh." *Field Columbian Museum Publications*, Geological Series 2, no. 4 (August 1903): 165–196.

Tschopp, Emanuel, Octávio Mateus, and Roger B. J. Benson. "A Specimen-Level Phylogenetic Analysis and Taxonomic Revision of Diplodocidae (Dinosauria, Sauropoda)." PeerJ, 3, entry 857 (April 2015). www.peerj.com/articles/857.

New York Times. "Yale Brontosaurus Gets Head on Right at Last." October 26, 1981.

For Loree, one of my favorite vertebrates
—S. A.

To our curious explorers: Artemis, Mona, Niko, and Mallory
—C. H.

Special thanks to my expert readers.—S. A.
Dr. Michael D. D'Emic, associate professor, Department of Biology, Adelphi University
Dr. Sterling Nesbitt, paleobiologist, Virginia Tech
Dr. Emanuel Tschopp, vertebrate paleontologist and sauropod specialist, Freie Universität Berlin

Charlesbridge, 9 Galen Street, Watertown, MA 02472
www.charlesbridge.com

Printed in China • OPIC

The authorized representative in the EU for product safety and compliance is eucomply OÜ, Pärnu mnt 139b-14, 11317 Tallinn, Estonia, hello@eucompliancepartner.com, +33757690241

(hc) 10 9 8 7 6 5 4 3 2 1

Illustrations created digitally
Text type set in Grenadine MVB by Markanna Studios Inc.
Edited by Karen Boss
Designed by Diane M. Earley
Production supervised by Jennifer Most Delaney

Library of Congress Cataloging-in-Publication Data
Names: Albee, Sarah author | Hsu, Chris illustrator
Title: Brontosaurus is back!: a dinosaur's name and return to fame / Sarah Albee; illustrated by Chris Hsu.
Description: Watertown, MA: Charlesbridge, 2026. | Includes bibliographical references. | Audience: Ages 6–9. | Audience: Grades K–1. | Summary: "Brontosaurus has quite a history, unearthed in a humorous and factual way from way back to present day."—Provided by publisher.
Identifiers: LCCN 2025029569 (print) | LCCN 2025029570 (ebook) | ISBN 9781623546366 hardcover | ISBN 9781632892836 ebook
Subjects: LCSH: Apatosaurus—Juvenile literature | Dinosaurs—Nomenclature—Juvenile literature | Discoveries in science—Juvenile literature | Paleontology—Mesozoic—Juvenile literature
Classification: LCC QE861.5.A385 2026 (print) | LCC QE861.5 (ebook) | DDC 567.913/8—dc23/eng/20251120
LC record available at https://lccn.loc.gov/2025029569
LC ebook record available at https://lccn.loc.gov/2025029570